Best
Blueberries

Lowfat Recipes

By
Sherri Eldridge

Illustrations by
Rob Groves

Best of Blueberries

Published by:
Harvest Hill Press
P.O. Box 55, Salisbury Cove, Maine 04672
207-288-8900
www.harvesthillpress.com

ISBN: 978-1-886862-14-2

Eleventh printing: June 2010

Printed in the U.S.A.
on certified environmentally-friendly
Acid-Free and Chlorine-Free paper

SUSTAINABLE
FORESTRY
INITIATIVE
Certified Fiber Sourcing
www.sfiprogram.org

The recipes in this book were created with the goal of reducing fat, calories, cholesterol and sodium. They also present a variety of fresh healthy foods, to be prepared with love and eaten with pleasure.

Credits:

Cover: Cotton print border gratefully used as a courtesy of:
Hoffman International Fabrics

Cover Design, Layout and Typesetting: Sherri Eldridge

Front Cover Watercolor and Text Line Art: Robert Groves

Text Typesetting and Proofreading: Bill Eldridge

PREFACE

The rich sweetness of blueberries is enjoyed by people throughout much of the northern hemisphere and, through exports, Europe.

There are two types of blueberries, the lowbush and highbush varieties. The fruit sizes differ for the two varieties. The lowbush fruit is small and generally considered to have a more distinct taste. The highbush has larger fruit, but a longer season. Both are excellent for eating fresh by the handful, preserving in delicious blueberry jam, or baked into pies, muffins and breads.

The lowbush blueberry grows in sandy acid soil throughout New England and northeastern Canada. These rocky "barrens" can support little other growth, but offer conditions favored by these hardy plants. The lowbush variety is usually referred to as the "wild" blueberry. The major commercial sources are Maine and the Canadian Maritimes. In both locations it is a major crop.

The highbush blueberry was first cultivated in the early 1900s when Elizabeth White of Whitesbog, New Jersey began collecting blueberry bushes that grew the largest fruit. Today, there are hundreds of varieties of highbush or "cultivated" blueberries. These beautiful plants are suitable for home landscaping and providing their wonderful fruit. The major commercial sources of highbush blueberries are New Jersey and Michigan, but the sandy soils of Oregon, Washington, British Columbia and North Carolina provide additional sources.

CONTENTS

Best of Blueberries
Lowfat Recipes

Bill's Best Blueberry Pancakes

2 cups cake flour
1 tablespoon baking powder
3 tablespoons sugar
1 egg
1 egg white
2 cups skim milk
2 tablespoons canola oil
1½ cups blueberries tossed
   in 2 tablespoons flour

Serving: 3 Pancakes
Protein: 8 gm
Carbs: 46 gm
Sodium: 265 mg

Calories: 270
Fat: 6 gm
Cholesterol: 37 mg
Calcium: 193 mg

SERVES 6

Sift together dry ingredients. In a separate bowl, beat eggs, then mix in milk and oil. Lightly stir the liquid mixture into the dry.

If fresh blueberries are not available, drain canned blueberries or defrost frozen berries in sieve, reserving blueberry juice for Blueberry Maple Syrup. Lightly fold drained blueberries into batter.

Cook on hot griddle sprayed with nonstick oil. Flip to cook on other side when light golden brown.

Blueberry and Pear Crepes

1¾ cups skim milk
1½ cups cold water
2 cups unbleached flour
1 egg
3 egg whites
2 tablespoons canola oil
2 cups blueberries
1 tablespoon cornstarch
¾ cup sugar
2 ripe pears, peeled and
    finely chopped
1 teaspoon vanilla
½ teaspoon lemon juice

To Serve: Roll crepes with
2 tablespoons filling. Dust with
powdered sugar.

Serving: 2 Crepes
Protein: 15 gm
Carbs: 118 gm
Sodium: 121 mg

Calories: 613
Fat: 9.5 gm
Cholesterol: 55 mg
Calcium: 166 mg

MAKES EIGHT 8-INCH CREPES

Blend milk and water into flour until perfectly smooth, leaving no lumps. Whisk in eggs and oil. Cover and refrigerate 30-60 minutes.

Mash 4 tablespoons fresh blueberries in a sieve, or drain canned or frozen blueberries, reserving ¼ cup of juice. Mix juice with cornstarch, sugar and pears and cook over medium-low heat until thickened. Mix in vanilla, lemon juice and blueberries. Keep warm.

Preheat crepe pan on medium-high. Spray hot pan with nonstick oil. Quickly pour ½ cup batter in middle of pan, tilt to cover. Crepe will be lightly browned in 30 seconds. Flip, and cook 20 seconds on the other side. Turn out each crepe onto a cloth towel. Do not stack. Roll crepes with filling, serve warm.

Blueberry Buttermilk Waffles

2 cups cake flour
1 tablespoon baking powder
1¾ cups lowfat buttermilk
1 egg, separated
2 egg whites
2 tablespoons canola oil
2 tablespoons sugar
1¼ cups blueberries

Serving: 1 Waffle
Protein: 7 gm
Carbs: 41 gm
Sodium: 153 mg

Calories: 240
Fat: 6 gm
Cholesterol: 35 mg
Calcium: 33 mg

MAKES 6 WAFFLES

Heat waffle iron while mixing batter.

Sift together flour and baking powder. Slowly stir in buttermilk, beaten egg yolk and oil. Beat egg whites with sugar, then fold into batter. Stir in blueberries.

Spray griddle with nonstick oil and ladle batter onto hot waffle iron. Cook until light golden brown. Repeat process for each waffle.

Blueberry Syrup

1 cup maple syrup
½ cup blueberry juice or
    juice pressed from berries

Serving: 2 Tablespoons   Calories: 84
Protein: 0 gm   Fat: 0 gm
Carbs: 22 gm   Cholesterol: 0 mg
Sodium: 5 mg   Calcium: 33 mg

MAKES 1¼ CUPS

Combine maple syrup and blueberry juice in medium saucepan. Rapidly boil down over medium-high heat for 25 minutes. Serve hot.

Blueberry Crumble

2 cups blueberries
½ cup sugar
juice of 1 lemon
1½ tablespoons canola oil
2 tablespoons apple juice
    concentrate
3 tablespoons brown sugar
1 cup unbleached flour

SERVES 4

Put washed berries in small shallow baking dish. Sprinkle with sugar and lemon juice.

Blend oil, apple juice concentrate, brown sugar and flour. Sprinkle over berries. Bake in 350° oven for 25 minutes. Serve warm.

Serving: 1/4 Recipe
Protein: 3 gm
Carbs: 57 gm
Sodium: 11 mg

Calories: 280
Fat: 5.5 gm
Cholesterol: 0 mg
Calcium: 19 mg

Blueberries in the Garden

Blueberries are easy for the home gardener to grow. The simplest way to start a blueberry garden is to buy plants and place them in a prepared bed. Highbush blueberries can also be started from root cuttings, but will need 4-5 years to become productive. Lowbush varieties, which are wild to begin with, require only a chunk of blueberry sod from an existing patch. The roots and rhizomes send up new stems, and the patch will spread and grow.

Both the highbush and lowbush blueberry plant require an acid, well-drained soil, good sunlight and adequate moisture. Be sure to have at least two varieties, three is a good idea. A self-pollinating bush will usually produce fewer and smaller berries.

Blueberry bushes are easy to care for. Highbushes need only a little pruning after their first 4-5 years, in late fall or early winter. Lowbush varieties are burned back every 2-3 years, then send out new growth.

The most annoying pest for blueberry gardeners are birds, which like the sweet berries as much as we do. If you can keep the birds away, the sweet blueberries will be well worth the care.

Blueberry Whip

1 cup blueberries
1 cup nonfat vanilla yogurt
½ cup sugar
2 egg whites

Serving: 2 Tablespoons
Protein: 1 gm
Carbs: 7 gm
Sodium: 11 mg

Calories: 30
Fat: 0 gm
Cholesterol: 0 mg
Calcium: 15 mg

MAKES 3 CUPS

Gently fold blueberries into yogurt.

In a separate bowl gradually add sugar while beating egg whites until stiff. Fold egg whites into yogurt. Serve at once over waffles, pancakes or blueberry pie.

Very Blueberry Muffins

½ cup nonfat plain yogurt
½ cup lowfat cottage cheese
1 egg
2 egg whites
1 cup sugar
2 tablespoons canola oil
1 tablespoon lemon juice
1 teaspoon vanilla extract
2 cups unbleached flour
1 tablespoon baking powder
½ teaspoon baking soda
1¼ cups blueberries dusted
   with 2 tablespoons flour

Serving: 1 Muffin
Protein: 6 gm
Carbs: 37 gm
Sodium: 194 mg

Calories: 199
Fat: 3 gm
Cholesterol: 19 mg
Calcium: 69 mg

MAKES 12 MUFFINS

Preheat oven to 400°. In a large mixing bowl, beat yogurt and cottage cheese with electric beater on high speed for 3 minutes. Add eggs, sugar, oil, lemon juice and vanilla. Beat well. Slowly add flour, baking powder and baking soda. Beat for 3 minutes. By hand, gently fold in dusted blueberries.

Lightly spray muffin tins with nonstick oil. Fill tins ¾ full. Bake 25 minutes, or until a toothpick inserted in center of muffin comes out clean.

Blueberry Corn Muffins

1 cup blueberries
2 tablespoons flour
1 egg
½ cup packed brown sugar
2 tablespoons canola oil
½ cup skim milk
½ teaspoon vanilla
1 cup yellow cornmeal
1 tablespoon baking powder
1 cup unbleached flour

Serving: 1 Muffin	Calories: 156
Protein: 3 gm	Fat: 3 gm
Carbs: 29 gm	Cholesterol: 18 mg
Sodium: 215 mg	Calcium: 106 mg

MAKES 12 MUFFINS

Preheat oven to 400°. Spray muffin tins with nonstick oil. If using canned or frozen blueberries, drain well.

Toss blueberries in 2 tablespoons flour to coat.

In a mixing bowl, beat egg, brown sugar, oil, milk and vanilla. With a few swift strokes, mix in remaining ingredients. Gently fold blueberries into batter.

Fill muffin tins ¾ full. Bake for 25 minutes or until toothpick inserted in muffins comes out clean. Serve warm.

Quick Blueberry Tea Bread

4 tablespoons canola oil
1½ cups sugar
2 eggs
2 egg whites
4 cups unbleached flour
1 tablespoon baking powder
½ teaspoon salt
1½ cups skim milk
2 cups blueberries (if frozen,
    thaw and pat dry)

Serving: 1 Slice
Protein: 4 gm
Carbs: 31 gm
Sodium: 268 mg

Calories: 164
Fat: 3 gm
Cholesterol: 18 mg
Calcium: 105 mg

MAKES 2 LOAVES

Preheat oven to 350°. Spray two 9" x 5" loaf pans with nonstick oil.

Using electric mixer, cream oil and sugar in a large bowl. Add eggs 1 at a time, beating well after each addition. In a separate bowl, combine flour, baking powder and salt. Add dry ingredients to egg mixture alternately with milk, beginning with dry ingredients. Gently fold blueberries into batter.

Divide batter between prepared pans. Bake until tester inserted in center comes out clean, about 50 minutes.

Cool breads in pans 20 minutes, then turn out onto rack. Cut each loaf into 12 slices. Serve with warm honey, cinnamon and a hot cup of tea.

Blueberries for Health

Native North Americans used blueberries for many purposes. They believed blueberries had magical powers and prized the berries for their healing abilities. Strong blueberry tea relieves pain, and both blueberry juice and its syrup can quiet coughs.

For years, people have known the blueberry was nutritious as well as delicious. Analysis has boosted that knowledge by showing exactly what blueberries are made of. Blueberries have no fat, cholesterol, or sodium. One cup of blueberries supplies about a third of the daily vitamin C requirement for adults. This same cup of blueberries also provides 85 milligrams potassium, 18 grams carbohydrates, 4 grams of fiber and only 80 calories.

Recent studies show that the pigment in blueberry skins has extra health benefits. The blue pigment, an anthocyanin, is a potent antioxidant in the flavonoid group of compounds. Anthocyanins are thought to help control diabetes, improve blood circulation, reduce eyestrain, prevent cancer, retard the effects of aging and improve memory skills. The United States Department of Agriculture recently rated 40 common fruits and vegetables for their disease-fighting antioxidant capacity, and blueberries rated number one!

Blueberry Kuchen with Almond Topping

Crust and Fruit:
1¼ cups unbleached flour
¼ cup sugar
½ teaspoon baking powder
¼ cup canola oil
1 egg
2 cups fresh blueberries
    tossed in ¼ cup flour

Almond Topping:
12 oz. nonfat cream
    cheese, softened
½ cup packed brown sugar
1 teaspoon almond extract
1 egg
1 egg white
3 tablespoons skim milk
1 cup unbleached flour
1 teaspoon baking powder
6 tablespoons sliced almonds

Serving: 1 Piece	Calories: 200
Protein: 7 gm	Fat: 6 gm
Carbs: 30 gm	Cholesterol: 28 mg
Sodium: 171 mg	Calcium: 80 mg

MAKES 16-PIECE KUCHEN

Preheat oven to 350°. Spray a 9" x 12" pan with nonstick oil.

Crust and Fruit: Combine flour, sugar and baking powder. Work in oil with fingertips or pastry cutter. Quickly blend in egg. Distribute the dough in dabs over the bottom of pan. Freeze 15 minutes. Flour fingertips and press dough over bottom and up sides. Spread blueberries over dough. Bake 15 minutes.

Topping: With electric beater, combine cream cheese, brown sugar and almond extract until fluffy. Add eggs and milk. In a separate bowl, sift flour and baking powder, then add almonds. Combine mixtures. Spread topping over baked blueberries, bake 20 minutes more.

Blueberry Peach Parfait

1½ lbs. peaches
3 tablespoons apple juice
    concentrate
¼ cup packed brown sugar
1 teaspoon cinnamon
2 pints blueberries
4 cups nonfat vanilla yogurt
6 sprigs of fresh mint

Serving: 1 Parfait
Protein: 9 gm
Carbs: 67 gm
Sodium: 106 mg

Calories: 295
Fat: 0.5 gm
Cholesterol: 0 mg
Calcium: 269 mg

SERVES 6

Concasse peaches: score an X in bottom and top of peaches. Place in boiling water for 4 minutes, then remove and place in ice water for 2 minutes. Skins will now easily peel off.

Chop peaches into half-inch pieces. Place in saucepan with apple juice concentrate, brown sugar and cinnamon. Cook on low heat for 15 minutes, stirring occasionally. Process half of the peaches in blender until smooth. Combine both peach mixtures, chill.

Assemble parfaits: use six 12-oz. parfait glasses or tumblers. Place ¼ cup berries in each glass, top with ¼ cup peaches, then ¼ cup yogurt. Repeat process for a second layer. Top with blueberries, garnish with fresh mint sprigs.

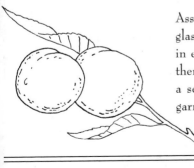

Blueberry BBQ Sauce

1 tablespoon olive oil
1 yellow onion, diced
3 cloves crushed garlic
¼ cup dry red wine
½ cup cider vinegar
5 cups fresh or frozen
    blueberries
1 cup ketchup
¼ cup packed brown sugar
2 tablespoons
    Worcestershire sauce
juice of 1 lemon
pinch of cayenne
pinch of chili powder
pinch of salt
pinch of pepper

Serving: 2 Tablespoons
Protein: 0 gm
Carbs: 8 gm
Sodium: 17 mg

Calories: 37
Fat: 0.5 gm
Cholesterol: 0 mg
Calcium: 7 mg

MAKES 1 QUART

In a large saucepan or stock pot, heat oil over medium heat. Add onion and garlic, and sauté until translucent but not brown. Add wine and vinegar and bring ingredients to a simmer. Stir in blueberries, ketchup, brown sugar, Worcestershire sauce, lemon juice and spices. Bring the mixture to a boil. Simmer over low heat until thick, about 2 hours.

When the sauce has thickened, purée in a food processor or blender. Taste, and adjust seasonings.

This sauce is an excellent baste on chicken and fish, and is especially good on a barbecue.

Blueberry Stuffing

SERVES 10

1½ cups diced yellow onion
¾ cup diced celery
½ cup diced carrots
1 tablespoon olive oil
½ cup chopped nut meats
2½ cups vegetable broth
2 tablespoons parsley
¼ teaspoon thyme
½ teaspoon chopped sage
½ teaspoon white pepper
5 cups diced stale bread
1½ cups blueberries

Sauté onion, celery and carrots in olive oil until onion is clear. Add chopped chestnut meat and broth, simmer for 30 minutes. Mix in spices and diced bread, stir well. Gently fold in blueberries. Cover and let rest 15 minutes before using stuffing, or bake in covered casserole at 325° for 1 hour.

Serving: 1/10 Recipe
Protein: 4 gm
Carbs: 31 gm
Sodium: 220 mg

Calories: 166
Fat: 3 gm
Cholesterol: 0 mg
Calcium: 65 mg

Blueberry Relish

3 cups blueberries
3 cooking apples, peeled,
    cored and chopped
1 cup sugar
½ teaspoon allspice
½ teaspoon nutmeg
½ teaspoon mace

Serving: 2 Tablespoons   Calories: 26
Protein: 0 gm   Fat: 0 gm
Carbs: 7 gm   Cholesterol: 0 mg
Sodium: 1 mg   Calcium: 1 mg

MAKES ABOUT 3 PINTS

Wash and drain blueberries, set aside.

Combine remaining ingredients in a large pot and bring to a boil. Lower heat and simmer until thickened. Remove from heat and stir in blueberries. Cover and let rest 10 minutes. Spoon into hot sterilized jars. Relish will continue to thicken as it cools. Will stay fresh in refrigerator for about 3 weeks.

Blueberry Sauce

½ cup sugar
½ cup water
pinch of salt
2 tablespoons cornstarch
2 cups blueberries
1 tablespoon lemon juice
1 teaspoon grated lemon
    peel

Serving: 2 Tablespoons   Calories: 38
Protein: 0 gm   Fat: 0 gm
Carbs: 10 gm   Cholesterol: 0 mg
Sodium: 9 mg   Calcium: 2 mg

MAKES ABOUT 2 CUPS

Combine sugar, water, salt and cornstarch. Cook, stirring frequently until mixture boils and thickens. Add blueberries, bring to a boil. Simmer 5 minutes. Stir in lemon juice and peel. Let sauce cool slightly before using. Serve over ice cream, vanilla yogurt, or fresh melon.

Using Blueberries

Who can resist the temptation of grabbing a handful of fresh blueberries from the basket? And, fresh frozen, blueberries will retain their flavor and texture up to two years. Fresh, frozen or preserved, the tasty blueberry is a very versatile fruit in the kitchen.

Blueberry pie is a classic irresistible dessert or breakfast pie. It can be served with lowfat ice cream, a dollop of fresh yogurt or nonfat whipped topping. For variety, mix grated apple, crushed pineapple or frozen orange juice concentrate into the berries. Vanilla and lemon juice also complement their flavor.

Blueberries are very popular in muffins, quick breads and pancakes. Vary the flour mixture to allow for some oatmeal, bran, or whole wheat flour. Pecans are wonderful in blueberry baked goods, but other nut meats can provide new flavor combinations. An overly ripe mashed banana is a nice addition to the batter of a batch of beautiful blueberry muffins.

Homemade blueberry jams and jellies taste as fresh as the day they were made. Blueberries have also made their way into chutneys, relishes, glazes, ice cream, vinaigrettes and wines.

Perfect Blueberry Pie

Crust:

2½ cups unbleached flour

3 tablespoons sugar

4 tablespoons canola oil

5 tablespoons ice water

4 tablespoons cold skim milk

Blueberry Filling:

2 tablespoons lemon juice

4 cups wild blueberries, dusted with ¼ cup flour

½ cup sugar

¼ cup packed brown sugar

1 tablespoon quick tapioca

½ teaspoon cinnamon

½ teaspoon nutmeg

Glaze:

1 tablespoon skim milk

1 tablespoon sugar

Serving: 1 Piece	Calories: 358
Protein: 5 gm	Fat: 7.5 gm
Carbs: 68 gm	Cholesterol: 0 mg
Sodium: 14 mg	Calcium: 34 mg

MAKES ONE 8-PIECE PIE

Preheat oven to 400°. Spray 9-inch pie plate with nonstick oil.

Combine flour and sugar. With pastry cutter or knives, cut oil into mixture. Use fork to blend in water and milk. Divide dough in half, and roll each between two sheets of wax paper. Place bottom crust in pie plate.

Sprinkle lemon juice over blueberries. In a separate bowl, mix sugars, tapioca, cinnamon and nutmeg. Fold blueberries into sugar mixture. Let stand 45 minutes, then gently stir. Pour into pie shell. Cover with second crust, crimp edges with fingers, flute and make 5 small slits in top. Brush top with milk, sprinkle with sugar. Bake 10 minutes at 400° then reduce heat to 350° and bake 25 minutes more.

Pineapple-Blueberry Crumb Pie

MAKES ONE 8-PIECE PIE

Pie Crust:
1½ cups unbleached flour
2 tablespoons sugar
4 tablespoons canola oil
6 tablespoons skim milk
2 tablespoons flour
¼ cup packed brown sugar

Filling:
1½ cups crushed pineapple
3 cups blueberries,
   tossed in ¼ cup flour
1 teaspoon cinnamon
¼ cup sugar

Topping:
¼ cup chopped pecans
½ cup unbleached flour
½ cup packed brown sugar
¼ cup sugar
2 teaspoons cinnamon
3 tablespoons pineapple juice

Serving: 1 Piece
Protein: 5 gm
Carbs: 73 gm
Sodium: 17 mg
Calories: 389
Fat: 9.5 gm
Cholesterol: 0 mg
Calcium: 59 mg

Preheat oven to 400°. Spray 9-inch pie plate with nonstick oil.

Crust: Combine flour with sugar. Use pastry cutter or knives to cut oil into flour. Blend in milk. Roll into ball, wrap in plastic and chill 15 minutes. Roll dough between two sheets of wax paper, place in pie plate. Mix the remaining 2 tablespoons flour and ¼ cup brown sugar, sprinkle over unbaked crust.

Press moisture from pineapple, reserving juice for topping. Combine all filling ingredients and spoon into pie crust. Blend topping ingredients until moisture is evenly distributed, crumble over pie.

Bake 15 minutes at 400°, reduce oven to 325° and bake 30 minutes more, or until topping is golden brown.

Blueberry Sour Cream Cheesecake

MAKES 8-PIECE CHEESECAKE

Cheesecake Crust:

2 cups lowfat graham
    cracker crumbs
2 tablespoons canola oil
2 tablespoons brandy

Cheesecake Filling:

1 egg
1 egg white
8 oz. softened nonfat cream
    cheese
1 cup lowfat sour cream
1 cup sugar
2 teaspoons vanilla

Blueberry Topping:

1 cup wild blueberries
2 teaspoons quick tapioca
1 teaspoon lemon juice
½ cup sugar

Combine crust ingredients in mixing bowl. Spray a 9-inch springform pan with nonstick oil. Press graham cracker mixture into pan and 2½ inches up sides. Bake in 375° oven 3 minutes, then chill.

Beat filling ingredients together. Pour into crust and bake 35 minutes at 375°. Chill well before serving.

Combine blueberries and tapioca in saucepan, crushing a few berries to release juice. Cook on low heat 5 minutes. Stir in lemon juice and sugar, then raise heat to medium. Simmer until thickened, about 15 minutes. Cool to room temperature. Spoon sauce over slices of chilled cheesecake on serving plates.

Serving: 1 Piece
Protein: 8 gm
Carbs: 61 gm
Sodium: 312 mg

Calories: 343
Fat: 6.5 gm
Cholesterol: 29 mg
Calcium: 82 mg

Blueberry Coffee Cake

Cake:
2 cups unbleached flour
½ cup sugar
1 tablespoon baking powder
½ teaspoon baking soda
1 teaspoon cinnamon
3 tablespoons canola oil
1½ cups lowfat buttermilk
1 egg
1 egg white
2 cups blueberries

Blended Nut Topping:
½ cup finely chopped nuts
½ cup brown sugar
1 teaspoon ground ginger

Serving: 1 Piece
Protein: 5 gm
Carbs: 39 gm
Sodium: 175 mg

Calories: 234
Fat: 7 gm
Cholesterol: 18 mg
Calcium: 70 mg

MAKES ONE 12-PIECE CAKE

Preheat oven to 350°. Spray 2-quart glass baking dish with nonstick oil.

Combine all cake ingredients, except the blueberries. Using electric mixer, beat 30 seconds on low speed. Scrape bowl and beat 2 more minutes on medium speed, scraping bowl frequently.

Pour half of the batter into prepared baking dish. Spread half the blueberries over batter, then half of the nut topping. Repeat layers with remaining batter, blueberries and nut topping. Bake for 45-50 minutes or until toothpick inserted near center comes out clean.

Blueberry Jelly

3 cups blueberries
1 teaspoon lemon juice
1 cup sugar
1 pouch (3 oz.) liquid
    fruit pectin

Serving: 2 Tablespoons   Calories: 83
Protein: 0 gm   Fat: 0 gm
Carbs: 21 gm   Cholesterol: 0 mg
Sodium: 4 mg   Calcium: 3 mg

MAKES 2 JELLY JARS

Crush blueberries in saucepan, add lemon juice and cover pan. Cook over medium heat 20 minutes. Remove from heat and strain through a jelly bag. Squeeze well.

Return juice to saucepan, add sugar and pectin. Bring to a boil, cook 3 minutes. Pour into hot sterilized jelly jars. Seal according to manufacturer's directions.

Blueberry Jam

4 cups blueberries,
    fresh or frozen, thawed
3 cups sugar
1 teaspoon lemon juice
1¾ ounces powdered pectin

Serving: 2 Tablespoons   Calories: 110
Protein: 0 gm   Fat: 0 gm
Carbs: 28 gm   Cholesterol: 0 mg
Sodium: 2 mg   Calcium: 2 mg

MAKES 3 CUPS

Wash and pick over blueberries. Crush 1 cup of the fruit to release juice, then put in large pan with remaining blueberries.

Simmer over moderate heat for 10 minutes. Add sugar, lemon juice and pectin. Stir occasionally, and cook over medium heat 20 minutes.

Ladle into hot sterilized jars. Seal with a two-piece metal screw-down lid according to manufacturer's directions.

Blueberry Wine

MAKES 4 QUARTS

4 quarts water
8 cups sugar
4 quarts blueberries
1 oz. yeast
1 slice whole wheat toast

Serving: 1/2 Cup
Protein: 1 gm
Carbs: 60 gm
Sodium: 9 mg

Calories: 235
Fat: 0.5 gm
Cholesterol: 0 mg
Calcium: 7 mg

Boil 2 quarts of water with the sugar. Pour hot water over blueberries, then add 2 quarts cold water.

Pour into a large plastic or glass container. Spread yeast on toast and soften with water. Float toast, yeast side down, on top of mixture. Cover and let rest 2 weeks, stirring up from the bottom once a day.

After 2 weeks, strain through several layers of cheesecloth and return liquid to cleaned crock. Let rest 2 days. Siphon from the top into clean bottles, cap lightly by just resting caps on bottles. When bubbling stops, tighten caps and wax over the tops. Leave 4 to 6 months, undisturbed, in a cool dark place. When its time for a special celebration, chill bottles and enjoy!